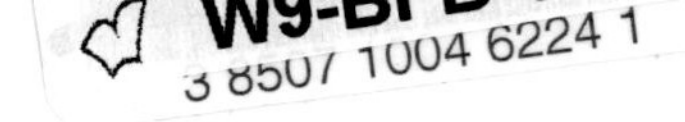

Unlocking the Secrets of Science

Profiling 20th Century Achievers in Science, Medicine, and Technology

Marc Andreessen and the Development of the Web Browser
Frederick Banting and the Discovery of Insulin
Jonas Salk and the Polio Vaccine
Wallace Carothers and the Story of DuPont Nylon
Tim Berners-Lee and the Development of the World Wide Web
Robert A. Weinberg and the Search for the Cause of Cancer
Alexander Fleming and the Story of Penicillin
Robert Goddard and the Liquid Rocket Engine
Oswald Avery and the Story of DNA
Edward Teller and the Development of the Hydrogen Bomb
Stephen Wozniak and the Story of Apple Computer
Barbara McClintock: Pioneering Geneticist
Wilhelm Roentgen and the Discovery of X Rays
Gerhard Domagk and the Discovery of Sulfa
Willem Kolff and the Invention of the Dialysis Machine
Robert Jarvik and the First Artificial Heart
Chester Carlson and the Development of Xerography
Joseph E. Murray and the Story of the First Human Kidney Transplant
Albert Einstein and the Theory of Relativity
Edward Roberts and the Story of the Personal Computer
Godfrey Hounsfield and the Invention of CAT Scans
Christaan Barnard and the Story of the First Successful Heart Transplant
Selman Waksman and the Discovery of Streptomycin
Paul Ehrlich and Modern Drug Development
Sally Ride: The Story of the First American Female in Space
Luis Alvarez and the Development of the Bubble Chamber
Jacques-Yves Cousteau: His Story Under the Sea
Francis Crick and James Watson: Pioneers in DNA Research
Raymond Damadian and the Development of the MRI
Linus Pauling and the Chemical Bond
Willem Einthoven and the Story of Electrocardiography
Edwin Hubble and the Theory of the Expanding Universe
Henry Ford and the Assembly Line
Enrico Fermi and the Nuclear Reactor
Otto Hahn and the Story of Nuclear Fission
Charles Richter and the Story of the Richter Scale
Philo T. Farnsworth and the Story of Television
John R. Pierce: Pioneer in Satellite Communications

Unlocking the Secrets of Science

Profiling 20th Century Achievers in Science, Medicine, and Technology

Godfrey Hounsfield and the Invention of CAT Scans

Susan Zannos

PO Box 619
Bear, Delaware 19701

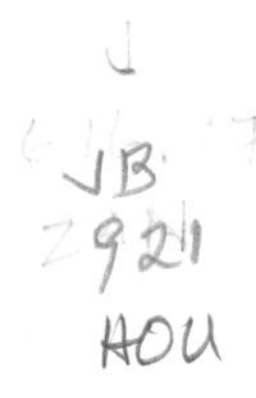

Unlocking the Secrets of Science

Profiling 20th Century Achievers in Science, Medicine, and Technology

Marc Andreessen and the Development of the Web Browser
Frederick Banting and the Discovery of Insulin
Jonas Salk and the Polio Vaccine
Wallace Carothers and the Story of DuPont Nylon
Tim Berners-Lee and the Development of the World Wide Web
Robert A. Weinberg and the Search for the Cause of Cancer
Alexander Fleming and the Story of Penicillin
Robert Goddard and the Liquid Rocket Engine
Oswald Avery and the Story of DNA
Edward Teller and the Development of the Hydrogen Bomb
Stephen Wozniak and the Story of Apple Computer
Barbara McClintock: Pioneering Geneticist
Wilhelm Roentgen and the Discovery of X Rays
Gerhard Domagk and the Discovery of Sulfa
Willem Kolff and the Invention of the Dialysis Machine
Robert Jarvik and the First Artificial Heart
Chester Carlson and the Development of Xerography
Joseph E. Murray and the Story of the First Human Kidney Transplant
Albert Einstein and the Theory of Relativity
Edward Roberts and the Story of the Personal Computer
Godfrey Hounsfield and the Invention of CAT Scans
Christaan Barnard and the Story of the First Successful Heart Transplant
Selman Waksman and the Discovery of Streptomycin
Paul Ehrlich and Modern Drug Development
Sally Ride: The Story of the First American Female in Space
Luis Alvarez and the Development of the Bubble Chamber
Jacques-Yves Cousteau: His Story Under the Sea
Francis Crick and James Watson: Pioneers in DNA Research
Raymond Damadian and the Development of the MRI
Linus Pauling and the Chemical Bond
Willem Einthoven and the Story of Electrocardiography
Edwin Hubble and the Theory of the Expanding Universe
Henry Ford and the Assembly Line
Enrico Fermi and the Nuclear Reactor
Otto Hahn and the Story of Nuclear Fission
Charles Richter and the Story of the Richter Scale
Philo T. Farnsworth and the Story of Television
John R. Pierce: Pioneer in Satellite Communications

Unlocking the Secrets of Science

Profiling 20th Century Achievers in Science, Medicine, and Technology

Godfrey Hounsfield and the Invention of CAT Scans

Printing 1 2 3 4 5 6 7 8 9

Library of Congress Cataloging-in-Publication Data

Zannos, Susan, 1934-

Godfrey Hounsfield and the invention of CAT scans/Susan Zannos.

p. cm. — (Unlocking the secrets of science)

Summary: Describes the life and career of Godfrey Hounsfield, the engineer who codeveloped computerized axial tomography (CAT), and shared the 1979 Nobel Prize in physiology and medicine.

Includes bibliographical references and index.

ISBN 1-58415-119-6 (library)

1. Hounsfield, Godfrey, 1919- —Juvenile literature. 2 Tomography—Juvenile literature. 3. Electric engineers—Great Britain—Biography—Juvenile literature. 4. Nobel Prizes—Juvenile literature. [1. Hounsfield, Godfrey. 1919- 2. Engineers. 3. Tomography. 4. Nobel Prizes—Biography.] I. Title. II. Series.

RC78.7.T6 Z66 2002

616.07'572—dc21

2002023657

ABOUT THE AUTHOR: Susan Zannos has been a lifelong educator, having taught at all levels, from preschool to college, in Mexico, Greece, Italy, Russia, and Lithuania, as well as in the United States. She has published a mystery *Trust the Liar* (Walker and Co.) and *Human Types: Essence and the Enneagram* was published by Samuel Weiser in 1997. She has written several books for children, including *Paula Abdul* and *Cesar Chavez* (Mitchell Lane). Susan lives in Oxnard, California.

PHOTO CREDITS: cover: Corbis; p. 6 Corbis; p. 8 AP Photo; p. 12 Science Photo Library; p. 15 AP Photo; p. 16 Bettmann/Corbis; p. 20 AP Photo; p. 22 Clinical Radiology Dept., Salisbury District Hospital/Science Photo Library; p. 24 AP; p. 26 Tufts University; p. 28 Michael Nicholson/Corbis; p. 31 Digital Collections and Archives/Tufts University; p. 33 AP Photo/Elise Amendola; p. 35 AP; p. 36 Science Photo Library; p. 39 Raymond Damadian; p. 41 top AP; bottom Science Photo Library

PUBLISHER'S NOTE: In selecting those persons to be profiled in this series, we first attempted to identify the most notable accomplishments of the 20th century in science, medicine, and technology. When we were done, we noted a serious deficiency in the inclusion of women. For the greater part of the 20th century science, medicine, and technology were male-dominated fields. In many cases, the contributions of women went unrecognized. Women have tried for years to be included in these areas, and in many cases, women worked side by side with men who took credit for their ideas and discoveries. Even as we move forward into the 21st century, we find women still sadly underrepresented. It is not an oversight, therefore, that we profiled mostly male achievers. Information simply does not exist to include a fair selection of women.

Contents

When Godfrey Hounsfield was a young man, he had fun thinking up new ideas and seeing if he could make them work. Hounsfield was later awarded the Nobel Prize in Physiology or Medicine in 1979 for his work on computerized axial tomography.

Chapter 1

Professor Roentgen's Amazing Rays

Like many other inventors and scientists, Sir Godfrey Hounsfield, who was awarded the Nobel prize in physiology or medicine in 1979 for developing computerized axial tomography, (also called computer assisted tomography) owes a part of his success to the great inventors who came before him. One of these was the man who received the first Nobel Prize in physics in 1901 for discovering X rays, Wilhelm Roentgen.

Roentgen was a German university professor. A man with restless energy and boundless curiosity, he constantly explored different areas in physics. Unlike many of his contemporaries who considered that everything worth knowing was already known, and who were content with studying existing scientific theory rather than conducting experiments, Roentgen spent most of his free time in the basement laboratory of his home in Würzburg, Germany. On the night of November 8, 1895, Professor Roentgen was experimenting with passing an electric current through a vacuum tube when he became fascinated with a greenish glow that appeared on a screen covered with fluorescent material. He wanted to know what caused the strange glow.

He put a black cardboard box over the tube (called a Crookes tube), completely darkened the laboratory, and found that the rays passed through the box and continued to cause the green glow on the screen, which had been treated with a chemical that was sensitive to light. He called the rays X rays because *x* is the mathematical symbol used for the unknown. He soon found that the rays caused

Wilhelm Conrad Roentgen, who discovered x rays, is shown in this undated photograph. The x ray remains one of the greatest advances in the history of medicine.

photographic plates to fog over, just as they did when exposed to light. Then he realized that a plate could not even be kept in the same room with the X rays without fogging over. He took a photographic plate from his desk drawer and found that it had fogged over.

Instead of throwing the plate out, he developed it. To his considerable surprise he found the image of a key on the photographic plate. There had been no key in the drawer with the plate. But there had been a key on the top of the desk. The only logical explanation was that the mysterious rays had gone through the wooden desk but for some reason had been blocked by the metal key.

There followed an intense period of experimenting with the rays. Roentgen spent almost every waking hour in his laboratory. He was so preoccupied that his wife, friends, and colleagues at the university feared something might be wrong and affecting his health. A friend of the Roentgens reported Anna Bertha Roentgen's distress: "Mrs. Roentgen said she had to go through several terrible days. Her husband came late to dinner and usually was in a very bad humor; he ate little, didn't talk at all, and returned to his laboratory immediately after eating. He didn't reply when asked what was the matter."

In his laboratory, Wilhelm Roentgen was conducting experiment after experiment, allowing the rays to travel through all manner of materials. He found that they went through a large book and through tin foil, but, oddly, not so easily through glass. How strange it was that a substance transparent to light should partly block these peculiar rays. He found that the rays passed through a piece of wood painted on both sides, but when the wood was held sideways, lines appeared on the screen. He found that various metals offered different amounts of resistance, and that the most opaque metal was lead. That explained why glass and paint were harder for the rays to penetrate: both contained lead.

The most astonishing result of his experiments happened when Roentgen held his own hand between the source of the rays and the screen. He was doing an experiment to confirm that a lead disk would block the rays. To do this he held the disk in his fingers between the Crookes tube (the source of the rays) and the screen. When he turned to look at the image on the screen, he was thunderstruck. He saw, as he expected to see, the clear outline of the lead disk. What astonished him was that he clearly saw the image of the bones inside the hand holding the disk. The X rays made it possible to see inside the body!

Roentgen continued to experiment, spending weeks in his laboratory and recording his experiments with many photographs of different objects. Finally he was ready to share his findings with the world. He called Anna Bertha to come down. He had her place her hand on a photographic plate and leave it there for several minutes while he exposed it to the rays. The result was the first X ray photograph of a human hand—and it gave poor Bertha quite a shock to see her bones, with her wedding ring apparently floating around very insubstantial flesh.

By January 1896, Wilhelm Roentgen was ready to reveal his amazing discovery to the scientific world. His only public demonstration was at the Würzburg Physico-Medical Society on January 23. He simply and clearly described his experiments and his observations, and then demonstrated how the rays penetrated paper, wood, and tin. He then

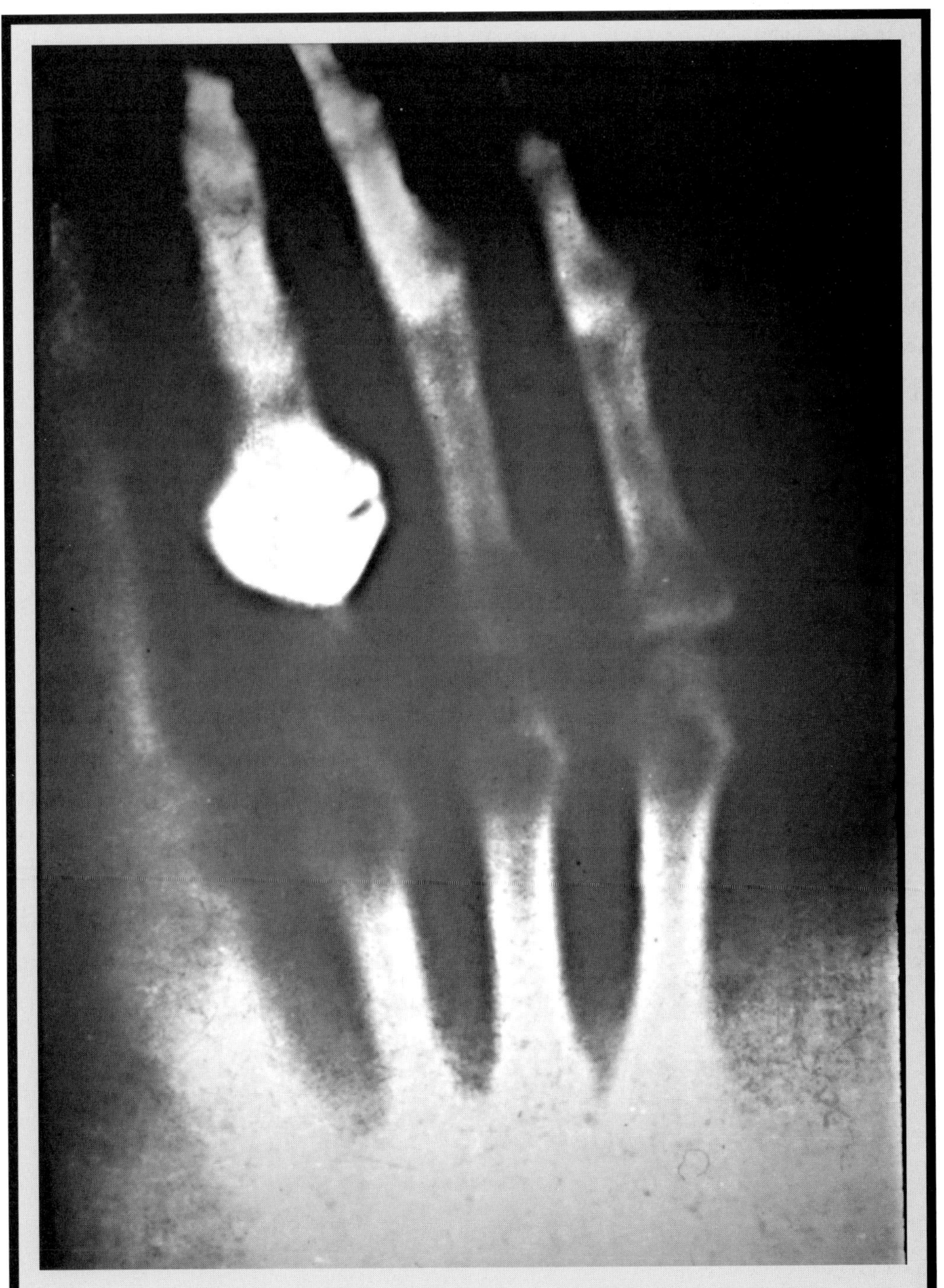

This is a reproduction of the original x ray that Roentgen took of his wife's hand. You can see where the ring sits on her finger.

showed the effect of the rays on his hand. Following his presentation there was enthusiastic applause from the audience of scientists and officials. Roentgen invited one of the scientists to have his hand photographed by the X rays, then developed the plate for the audience to see.

By the end of January the news about Professor Roentgen's mysterious rays had spread throughout the world. Other scientists immediately began studying and experimenting with them. By the end of February 1896, X rays were already being used by medical professionals. At Dartmouth College in Hanover, New Hampshire, physicist Edwin Frost made an X ray of a patient's fractured wrist for his brother, who was a doctor. In Montreal, Canada, X rays located a bullet in a gunshot victim's leg. The use of X rays for medical diagnoses immediately followed their discoveries.

Friends advised Roentgen to patent his discovery, but he had no interest in fame or wealth and was even annoyed by the disturbance the excitement of the scientific world caused in his life. Reporters came to interview him, and he got more letters than he could possibly answer. In an interview with a reporter from *McClure's Magazine,* published in April 1896, Roentgen was very open about his ignorance of the nature of the rays:

"It seemed at first a new kind of invisible light," Professor Roentgen said. "It was clearly something new, something unrecorded."

"Is it light?"

"No."

"Is it electricity?"

"Not in any known form."

"What is it?"

"I don't know."

During the first year scientists published hundreds of papers describing their studies of the rays. Roentgen himself carried out no further experiments. He had never been one to stay with a subject for long but always moved on to new areas of study.

What are Professor Roentgen's mysterious rays? In the century after the rays were first observed, scientists learned that they are related to visible light. X rays and visible light are both forms of radiation that move in waves, but X rays have very short wavelengths. (The wavelength is the distance between the tops of the waves.) In fact, the wavelengths of X rays are between 1/10 and 1/100,000 the wavelengths of visible light. It is their very, very short wavelength that allows the rays to pass through all but the densest materials.

The denser the material is, the fewer the number of rays that can pass through it. Just as a solid object will cast a shadow when sunlight hits it, dense materials create an X-ray shadow. This is why bones appear white on medical X-ray pictures. The dense material of calcium in the bones blocks the rays, so the rays do not get through to turn the photographic plate dark.

During the initial excitement about them, X rays were widely used for demonstrations and rather silly purposes. For a while—even into the 1950s—shoe stores had X-ray machines for customers. Customers could look through a special scope and see the bones of their feet inside the shoes to tell if the shoes fit properly. Unfortunately, many people who made demonstrations or who worked near X-ray machines suffered from prolonged exposure to X rays.

No one had realized that the rays could be harmful. In May 1896, at the annual National Electrical Exhibition in the United States, Thomas Edison arranged a public demonstration of the rays. Hundreds of people crowded through to see the bones in their hands, and the keys or metal objects they held, outlined on a screen. Because of the lengthy exposure to the rays, Edison's assistant developed severe burns and in 1904 finally died from radiation poisoning. Others also suffered from very painful burns, and many found that the penetrating rays could damage cells and cause cancer. Scientists learned to use lead shields and less powerful intensities of the X rays, and to limit exposure times to avoid injuries.

By the end of the 20th century, more than 200 million X-ray photographs were being taken every year to help with medical diagnoses of health problems. Doctors now use X rays not only to look at broken bones but also to locate tumors and other diseases of the internal organs. Chiropractors use X rays to reveal misalignments of the spine, and dentists take X rays of their patients' teeth to find out if there are any cavities.

People in the health care professions are not the only ones to use Professor Roentgen's no longer mysterious rays. Factories use X rays to inspect equipment for flaws. Engineers use X rays to check the structures of buildings. Archaeologists use X rays to view the contents of ancient

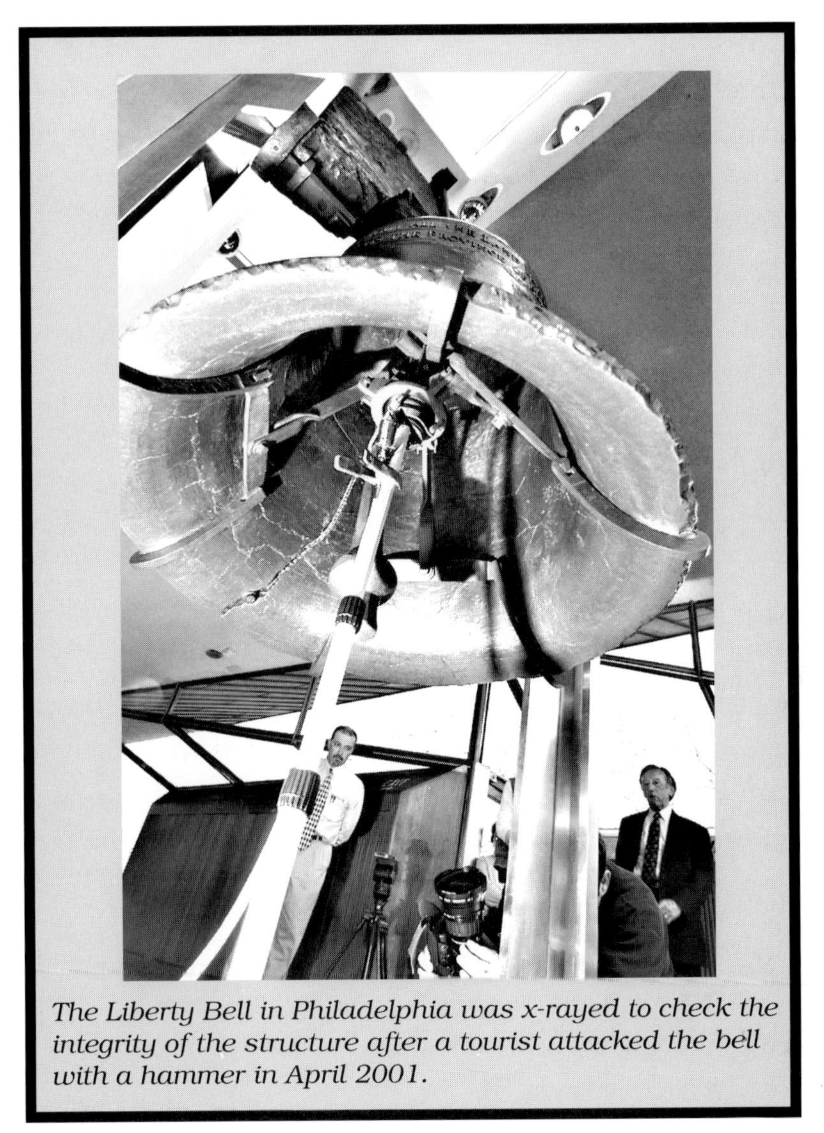

The Liberty Bell in Philadelphia was x-rayed to check the integrity of the structure after a tourist attacked the bell with a hammer in April 2001.

tombs. Museum curators use X rays to tell if old paintings are genuine or not.

But it was not until the development of the computer, and the genius of a man who realized that X-ray images could be stored and manipulated by computers, that X rays were used in an amazingly accurate diagnostic device known as the computerized axial tomography, or CAT, scanner. That man is Godfrey Hounsfield.

Godfrey Hounsfield had the idea that x rays could be controlled by computer to produce a much better view of the human body for medical purposes.

Chapter 2

How Things Work

Godfrey Newbold Hounsfield was born on August 28, 1919, in the village of Newark in Nottinghamshire, which is near the center of England. Thomas Hounsfield, Godfrey's father, had been an engineer, but at the end of World War I he bought a farm in Nottingham and became a farmer. Godfrey was the youngest of the five Hounsfield children, which was almost like being an only child since his two brothers and two sisters were quite a bit older and had their own more grown-up interests. "This gave me the advantage of not being expected to join in," he remembered in the autobiography he wrote for the Nobel Foundation Web page, "so I could go off and follow my own inclinations."

Young Godfrey's inclinations were to poke about the farm machinery and electrical equipment to see how things worked. Godfrey had a relative, Leslie Hounsfield, who was a well-known inventor and had designed an automobile: the Trojan, which had a two-stroke engine and had been unveiled in 1910. Whether or not Godfrey was inspired by his relative's success, he certainly followed in his footsteps.

His years between the ages of 11 and 18 were a time of experimentation. "In a village there are few distractions and no pressures to join in at a ball game or go to the cinema," he said. Far from regretting the lack of social life, he reveled in his experiments. He made an electrical recording machine. He constructed a glider and took flight from the tops of haystacks. Using acetylene, he launched tar barrels full of water "to see how high they could be waterjet propelled. It may now be a trick of the memory but I am sure that on one occasion I managed to get one to an altitude of 1000 feet!"

He found these projects far more interesting than the little country school he attended in the village of Newark, the Magnus Grammar School. He was not much of a scholar, although he did perk up a bit when studying mathematics and physics. These childhood years formed the pattern that Godfrey Hounsfield would follow throughout his life. He never did have much of an academic career, unlike nearly all the other Nobel Prize winners in scientific fields. He never got a Ph.D. or piled up a lot of academic credits. He was always too busy coming up with ideas and seeing if he could make them work.

Godfrey attended City and Guilds College in London in 1939, but when World War II began he immediately enlisted as a reservist in the Royal Air Force—his interest in flying had continued from his glider-and-haystack days. "I took the opportunity of studying the books which the RAF made available for Radio Mechanics," Hounsfield wrote in his autobiography. He took the test for that field and was given the position of Radar Mechanic Instructor in the RAF-occupied Royal College of Science, and later in the Cranwell Radar School. Under the pressures of wartime, the British were far more interested in finding people who could make things work, and who could explain to others how things worked, than they were in academic credentials.

The young Hounsfield was not content with merely teaching theory. He developed an oscilloscope, which is an instrument for visually demonstrating the changes in a varying electrical current, and other educational equipment. At the end of the war, in 1945, he received the Certificate of Merit for his work during the war. The Air Vice-Marshal appreciated Hounsfield's work and saw to it that he got a grant to continue his education. A year after he was discharged from the Royal Air Force, Hounsfield enrolled in the Faraday House Electrical Engineering College in London.

After he graduated in 1951, Hounsfield went to work for EMI, Electric and Musical Industries, Limited. This company was a worldwide conglomerate based in Britain. It was best known for producing records (and is still one of the world's most successful recording companies) and owned movie theaters, social clubs and dance halls. The entertainment business was not, however, why they had hired Godfrey Hounsfield.

EMI also did research and development of electronic products. Since they had almost unlimited funds to devote to such research, Hounsfield was able to pursue whatever interests seemed likely to produce commercially useful designs. At first he continued his work with radar and weapons guidance systems, which were of great importance during the Cold War years. Then he ran a small design laboratory, where, in the mid-1950s, he became interested in computers, which were at that time in the beginning stages of development. He designed various computer data storage systems, and in 1958 became the project engineer on a team that designed and constructed the first solid-state computer in Britain, the EMIDEC 1100.

"In those days the transistor was a relatively slow device," Hounsfield reported, "much slower than valves which were then used in most computers. However, I was

This is the corporate headquarters for EMI, Electric and Musical Industries, in Great Britain. EMI is best known for producing records, but they also do research and develop electronic products. Godfrey Hounsfield worked for EMI for many years. The Beatles also produced many records under the EMI label.

able to overcome this problem by driving the transistor with a magnetic core." His innovation increased the computer's speed and brought about the use of transistors in computing much sooner than had been expected. EMI sold a couple dozen large installations before the increasing speed of transistors made Hounsfield's method obsolete.

When his work with the EMIDEC 1100 was finished, Godfrey Hounsfield transferred to the EMI Central Research Laboratories, where he had full opportunity to explore whatever research he thought might eventually result in practical applications. This freedom to explore whatever interested him and try out new ideas suited him well, even though he admitted about his ideas that "99 percent of them turned out to be rubbish." One of the things he was working on was a large-scale memory system to increase the capacity of the EMI computers to have a one-million-word store, but the project turned out not to be commercially workable.

While he was doing research on computers that were capable of recognizing print, the first scanning devices, he wondered whether computers could also be programmed to recognize other kinds of patterns. If a computer could recognize printed characters, why couldn't it recognize other kinds of images? For example, why couldn't computers recognize X-ray images?

While doing research, Hounsfield came across the work of a German mathematician, J. Radon, who in 1917 had written an article describing a method for representing an object in space through mathematical equations. A series of pictures of the object would be taken from different angles—as though a photographer were walking around it taking snapshots—and a three-dimensional representation of the object could be reconstructed mathematically. What if, Hounsfield wondered, the pictures were X-ray images?

Conventional X-ray images were very useful for internal problems that were characterized by a contrasting density of tissues. They did a great job of showing bone fractures, and a good job of detecting lung cancer because the tumors were much denser than the surrounding tissues. But they could not show problems that involved tissues that were of similar density. In particular, X rays could give almost no help to doctors who needed to examine the human brain, first because the brain is encased in the dense bone of the skull, and second because all of the brain tissues are about the same density.

Godfrey Hounsfield figured that if he could take X rays of separate slices, or cross sections, and then have a computer reconstruct the layers, he could produce much greater accuracy in the images. Ordinary X rays could provide only one percent of information. "I devised a system that would use 100 percent," Hounsfield is quoted as saying in the April 1980 *Current Biography.*

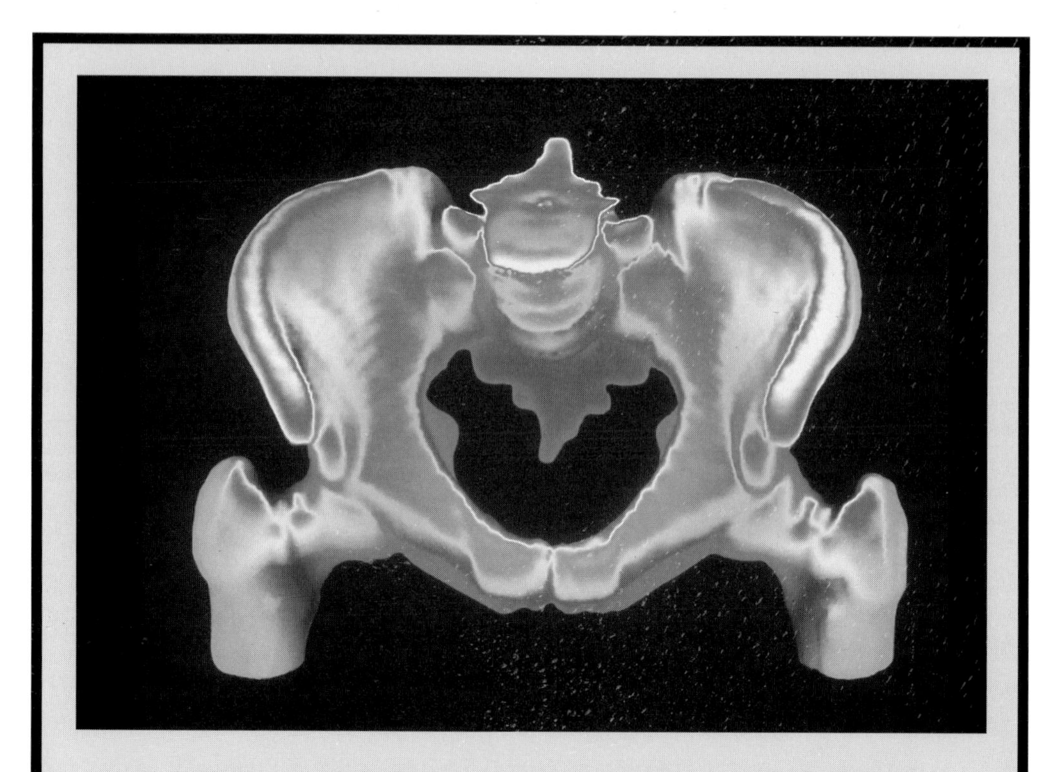

This is a three-dimensional computed tomography scan of bones of a normal adult pelvis. The pelvis supports and protects internal organs and tissues of the abdomen, and provides a site for the attachment of muscles of the trunk and lower limbs.

Hounsfield began working with his idea in 1967, but it was several years before computer technology became sophisticated enough to make it work. In the meantime he spoke with the British Department of Health and Social Services about his plan, and they were definitely interested. He also began working with radiologists James Ambrose and Louis Kreel for help with the X rays.

An interesting historical footnote to this period of time concerns a group of young musicians from Liverpool, the Beatles. They made their first recordings for EMI, and made quite a good amount of money. When Beatle Paul McCartney was deciding how to invest his money, he found out about the research Godfrey Hounsfield was doing in EMI's Central Research Laboratories and thought it would be a good choice for an investment that could benefit humanity. It was.

As always—from the time he was a curious boy taking apart machinery on the farm through his work with cutting-edge computer technology—Hounsfield's approach was hands-on and practical. The working model of a scanner that he constructed in the EMI laboratory produced very positive results when he did the first scans on preserved human brains, sundry parts of pigs, and the brains of freshly killed steers, which he transported personally through the London traffic.

"As might be expected," Hounsfield remembered in his autobiography, "the programme involved many frustrations, occasional awareness of achievement when particular technical hurdles were overcome, and some amusing incidents, not least the experiences of travelling across London by public transport carrying bullock's brains for

Paul McCartney, a former member of the musical group, the Beatles, is shown here in 1970 with his then-wife, Linda. The Beatles made several recordings for the EMI label and Paul invested his own earnings in Godfrey Hounsfield's research. Linda died from breast cancer in April 1998.

use in evaluation of an experimental scanner rig in the Laboratories."

The results Hounsfield got from his prototype scanner were positive, but they certainly weren't fast. It took nine days to take the pictures—an amount of time that was not important to a dead steer but would have seemed a bit much to a live human patient. Furthermore, the best available supercomputer available in 1967 took two hours to process the data.

Meanwhile, on the other side of the Atlantic Ocean, another man had come up with a very similar idea and had been working on it. Physicist Allan Cormack, who would share the 1979 Nobel Prize in physiology or medicine with Hounsfield, was approaching his research in a much more theoretical manner, but his conclusions were much the same as Hounsfield's more practical results would be.

This photograph of Allan Cormack was taken after he was notified about his Nobel Prize. He is shown with the aluminum and wood model he used to demonstrate his theories.

Chapter 3

Another Approach

Allan MacLeod Cormack was born in Johannesburg, South Africa, in 1924. His parents, a civil engineer and a teacher, had emigrated from Scotland to South Africa before World War I. When Allan was 12 years old, his father died and his mother took her three children to Cape Town. Allan, the youngest child, went to Rondebosch Boys High School, where he excelled in astronomy, physics and mathematics and was active in tennis and acting.

When he graduated from high school, Allan realized that even though astronomy was his favorite subject, the chances for making a living at it weren't good. He enrolled at the University of Cape Town to study electrical engineering, but before completing the course he changed to studying physics. He got his bachelor of science degree in physics in 1944 and his master's degree in 1945.

Like Godfrey Hounsfield, Allan Cormack never got a Ph.D. Instead he had the opportunity to work as a research student at the prestigious Cavendish Laboratory at Cambridge University in England. While there he worked with radioactive helium and attended lectures on quantum mechanics. He got rather more than he had bargained for in the quantum mechanics lectures—he met an American girl, Barbara Seavey, and a year and a half later he wanted to marry her, but he was broke.

On the basis of his work at Cambridge, Cormack was appointed as lecturer in physics at the University of Cape

This is an aerial view of some of the buildings at Cambridge University in England where Allan Cormack had the opportunity to work as a research student at their prestigious Cavendish Laboratory.

Town. He took his new wife back to South Africa with him. He also became a medical physicist at the Groote Schuur Hospital. Even though Cape Town did not have sophisticated laboratories like Cavendish, and though he felt isolated from other nuclear physicists, Cormack did have freedom to study. He published several scientific papers during this period.

It was in the radiology department of Groote Schuur Hospital that Cormack became involved with radiation treatment of cancer patients. Those experiences would eventually lead to the work for which he received the Nobel Prize that he shared with Godfrey Hounsfield. At the hospital, Cormack supervised the use of radioisotopes and calibrated the film badges that showed the amount of radiation the hospital workers received. He realized that it was necessary to have accurate information about how different body tissues absorb radiation. Only with such information could it be determined how to plan the radiation doses given to tumors. In thinking about how to get this information, he realized that there would also be diagnostic uses in locating tumors more accurately. Ordinary X-ray photographs were not precise enough.

The X-ray image could only show the total absorption along the path of the rays. It couldn't show how much was being absorbed by each of the tissues in line along the path. This caused particular difficulties with X-ray images of the head, where the skull absorbs most of the rays and makes it difficult for them to reach the soft brain tissue. Cormack figured out, just as Hounsfield had, that a series of measurements made with X-ray beams passing through at many different angles would reveal the different rates at which different parts of the brain absorbed the rays.

Where Hounsfield had set to work to solve the problems of how to gather the needed information with a working model, Cormack approached the same problems mathematically. He knew that multiple X rays would contain the necessary information. The problem was how to interpret the mass of data to reconstruct the interior details. He began by assuming that the X-ray beam would pass through the brain from many different angles but always in the same plane. This would result in a two-dimensional cross section. Then the process could be repeated in closely spaced parallel planes so that he could make a three-dimensional reconstruction.

This process of making thin slices of X-ray images is called tomography, from the Greeks words *tomos*, meaning "section," and *graphos*, meaning "writing." The technique eventually became known as computerized axial tomography, or CAT scanning. It is also called computed tomography, or CT scanning. Cormack continued for several years to work out the mathematical procedures for analyzing the X-ray data. He carried out this project in his spare time, while his career took him in somewhat different directions.

As he explained in his autobiography for the Nobel Foundation Web page, "It seemed only reasonable that since my wife had willingly come out to the wilds of Africa with me that I should go to the wilds of America with her. In addition the United States was a very good place to do research." Not many would consider Harvard to be "the wilds of America," but that's where the Cormack family ended up.

Cormack conducted research in the cyclotron laboratory at Harvard University, studying the interactions of protons and neutrons. He went back briefly to Cape Town and then returned to the United States to become a physics professor at Tufts University in Medford, Massachusetts. He became a U.S. citizen in 1966. While he was in Cape Town, and later back in the United States, he conducted experiments to test his mathematical theories. "My main

Allan Cormack became a physics professor at Tufts University in Medford, Massachussetts. He served as chairman of the department from 1968 to 1976.

interest for most of this time was in nuclear and particle physics," Cormack said, "and I pursued the CT-scanning problem only intermittently, when time permitted."

He constructed bodies of aluminum and wood so that they would have materials in them with different densities. He created a thin beam of gamma rays (which work the same way as X rays do), with a Geiger counter on the opposite side from the beam source. The source and detector were fixed, while the aluminum-and-wood cylinder was mounted on a platform that could be moved relative to the beam. The method worked. It even detected irregularities in the aluminum. Cormack repeated the experiment with a more complicated model: an aluminum case representing the skull, a plastic interior representing soft brain tissue, and two disks representing tumors. Again the experiment was successful.

In 1963 and 1964 Cormack published two papers about his mathematical procedure and experimental results. He hoped that radiological physicists would be interested and pursue the experiments to develop a way to use the procedure for medical diagnoses, "but as there was practically no response I continued my normal course of research and teaching," he said. Nonetheless, he had demonstrated that his method for showing interior details with different X-ray absorptions in cross sections of a body actually worked. It was only a laboratory demonstration, but already computers had been used to speed up the mathematical calculations. The results, however, were displayed as graphs rather than as photographic images.

At Tufts University, Cormack went back to the research in particle physics that he had been doing at Harvard. He became a full professor of physics and served as chairman of the department from 1968 to 1976. Until the end of the 1960s, he did not know about the work Godfrey Hounsfield was doing, and Hounsfield was unaware of Cormack's work. When Cormack did learn of the EMI scanner, he sent Hounsfield a note of congratulations. The two men did not meet each other before they shared the Nobel Prize in 1979.

"In the period 1970–72," Cormack said, "I became aware of a number of developments in, or related to CT-scanning, and since then I have devoted much of my time

This is a statue of Jumbo the elephant outside Barnum Hall on the campus of Tufts University. Phineas T. Barnum, a circus showman and legendary huckster, left the jumbo-sized legacy for the tiny university.

to these problems." Unlike Hounsfield, who remained a bachelor, Allan Cormack was a family man. He and his wife and their three children settled in Winchester, Massachusetts. According to his report he led a rather sedentary life except for a little swimming and sailing in the summers.

Meanwhile, back in England, Godfrey Hounsfield had installed the first CAT scanner in Atkinson Morley's Hospital in Wimbledon in 1971. In April 1972 the EMI Corporation announced that CAT scanners were in production. The first model, the EMI CT 1000, was shown in Chicago in November. Also in November, Hounsfield and radiologist James Ambrose described the invention at the annual conference of the British Institute of Radiology. In an article about the scanner that Hounsfield wrote for the December 1973 issue of the *British Journal of Radiology*, he said, "It is possible that this technique may open up a new chapter in X-ray diagnosis." That was a considerable understatement.

The apparatus that Hounsfield produced had four parts: an X-ray generator; the scanning unit, which had an X-ray tube and a detector opposite it; a viewing unit on which the scan appeared; and a computer to process the data. The patient would lie on a table with his or her head inside the machine. As Hounsfield described it, "The patient's head is placed within a rubber cap in a circular orifice around which the X-ray source and detectors rotate." The scanning unit rotated one degree at a time around the head, taking 160 separate readings at each position; 180 degrees were covered, so 28,800 readings, which were converted by the detector into electrical signals, were fed into the computer. The computer then combined all the signals and presented to the viewing unit a picture of the brain that revealed the differences in tissue density.

The medical community immediately recognized the diagnostic value of Hounsfield's scanner. Even though the machines were extremely expensive (the first models sold for $300,000, and the price soon climbed to nearly $1 million), EMI had no difficulty selling them. By 1975 EMI was ready to introduce full-body scanners.

Doctors and patients were enthusiastic about the new diagnostic device because it could frequently replace the need for exploratory surgery, which was extremely risky when performed on the brain. But not everyone shared this enthusiasm. Some people viewed the scanners negatively because they were so expensive. As more and more hospitals ordered the machines, irate consumer groups complained about the price and claimed that they were only expensive toys for doctors to play with.

It was in this atmosphere of controversy over whether the CAT scanners were indeed worth their considerable expense that the Nobel Prize committee made their decision to award the 1979 prize in the area of physiology or medicine to Hounsfield and Cormack. Since then CAT scanners have been vastly improved and are used all over the world.

This is the 1979 Nobel prize that was shared by Godfrey Hounsfield and Allan Cormack.

Godfrey Hounsfield, the British inventor of the Computed Tomography, or CT scanner used for medical diagnoses, is shown beside the prototype scanner he designed in 1972.

Chapter 4

Awards and Honors

Because his initial concern had been to develop a scanner that would gather accurate images of brain tissue, Godfrey Hounsfield's first working model of a CAT scanner was built to produce head images only. The first one developed in his lab at EMI took hours to get the data for one cross section, and it took days for the computer to reconstruct an image from the data. By 1976, however, Hounsfield and EMI were producing whole-body scanners. In the quarter century since that time, there have been great improvements in speed and image quality. The latest systems can scan an entire chest area—taking data from 40 slices—in 10 seconds.

As CAT scan times have improved, larger areas of the body can be scanned in less time. This means greater comfort for the patient, since he or she does not have to lie absolutely still for long periods of time. Since the chest can be scanned in 10 seconds, patients can hold their breath for that brief amount of time and the movement of breathing will not blur the image. And in the much shorter time there is a lot less exposure to the potentially harmful X rays.

Godfrey Hounsfield remained at EMI, where he became head of the Medical Systems Section in 1972 as the first CAT scans were being produced, and then senior staff scientist in 1977. He is quoted in a 1980 *Current Biography* article as saying that EMI was "a very relaxed place to do original thinking; and there's just enough pressure to produce a product. It enables one to come to a logical conclusion rather quicker than one would in academic life." If there is a sense in which he was "married to his work," it was a marriage made in heaven.

Although he never became a part of the academic community, he was appointed professional fellow in imaging sciences at the University of Manchester in 1978. He received many honorary degrees from universities all over the world. During the 1970s he received so many awards he had little time for anything other than accepting them.

Among his many honors were the MacRobert Award of the Fellowship of Engineering in 1972, the Barclay Prize of the British Institute of Radiology in 1974, the Albert Lasker Basic Medical Research Award in 1975, the Duddell Medal and Prize of the Institute of Physics in 1976, and the Gairdner Foundation International Award in 1976, for which he traveled to Canada. The culminating honor was receiving the Nobel Prize in physiology or medicine in 1979, which he shared with Allan Cormack. He was knighted in 1981, which for an Englishman may be as great an honor as the Nobel Prize, so he is Sir Godfrey Hounsfield now.

His work and research at EMI continued through all the furor of awards and honors. He continued refining and developing CAT scanning techniques and became interested in other types of imaging, particularly nuclear magnetic resonance. In 1972, the year the first commercial CAT scan was ready, Dr. Paul Lauterbur first described the basic MRI technique. In 1972, Dr. Raymond Damadian, another pioneer in MRI, applied for the first patent for scanning the human body by NMR and in 1974, he received the patent for the NMR scanner. The first magnetic resonance image

Dr. Raymond Damadian, shown here, was the first to patent an NMR scanner. Damadian, owner of the FONAR Company, is a pioneer in MRI technology.

appeared in 1973—at that time Dr. Lauterbur called the technique "zeugmatography," a word he made up from the Greek word *zeugma*, meaning "that which is used for joining," but the name didn't stick. After that the technique was called nuclear magnetic resonance imaging, but in the

1970s and '80s people were very nervous about anything called "nuclear," so that part of the name was dropped. Not until the 1980s were MRI systems approved by the Food and Drug Administration. After that their use spread rapidly until now there are more than 10,000 in use worldwide.

It is easy to see why Godfrey Hounsfield was interested in MRI. For one thing, the company for which he worked, EMI, was very interested in any technology that would end up with a marketable product. For another, the problem of using computers to compile the data into images was one that continued to intrigue him. Although the method of gathering the data was different, using magnetic energy and radio waves rather than X rays to create cross-sectional images of the body, many of the technical problems were the same.

The main part of the MRI system is a big tube-shaped magnet with a magnetic field 30,000 times stronger than the pull of gravity on earth. The patient lies on a moving table that slides into the tube. During the examination, a radio signal is turned on and off. The energy from this signal is absorbed or reflected by atoms in the body and measured by the scanner. The data is then fed to a computer that reconstructs the echoes of the radio waves into visual images, which provide very sharp contrasting details between different tissues with very similar densities, such as gray and white brain matter.

"The method of picture reconstruction," Hounsfield has explained about his work with imaging systems, "was obtained by commonsense practical steps. Most of the available mathematical methods at the time were of an idealized nature and rather impractical." This statement

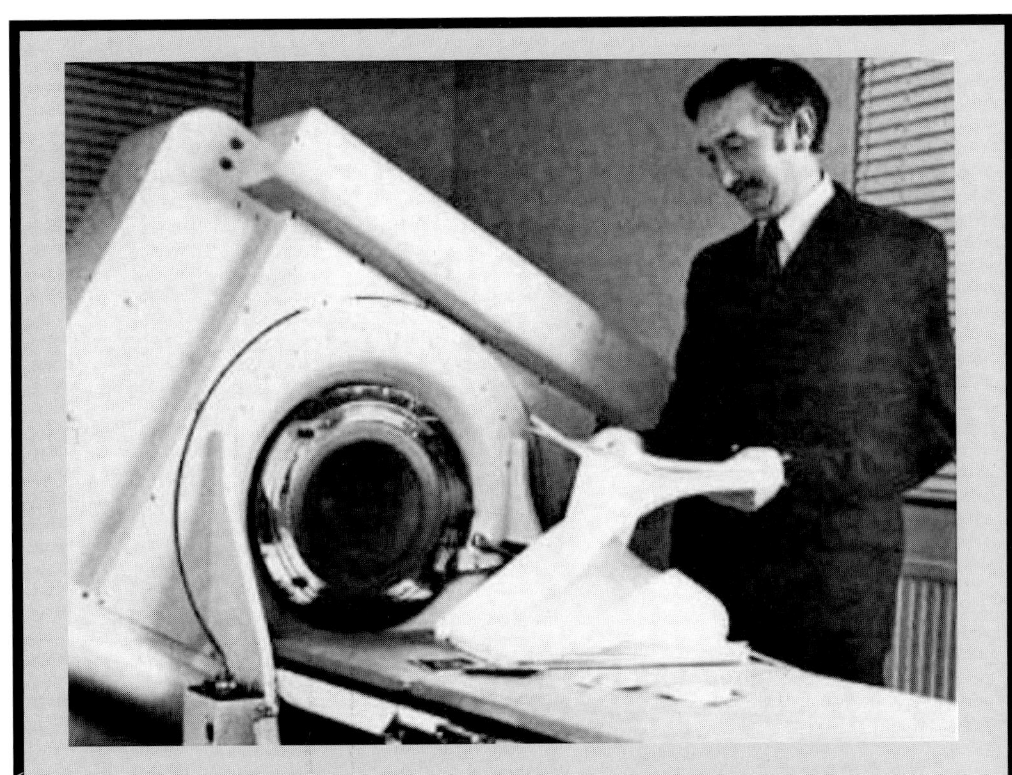

Godfrey Hounsfield is shown beside one of the first CAT scanners.

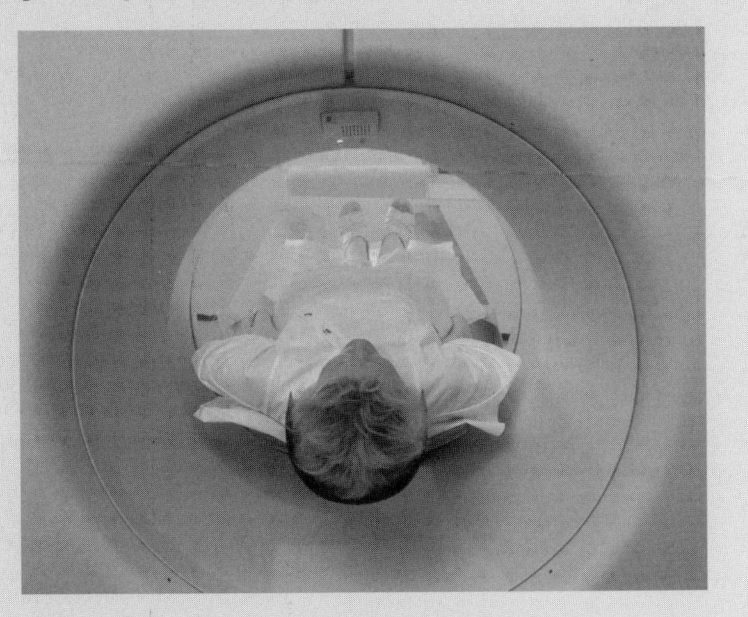

This is a photo of a female patient undergoing an upper-body CT scan.

seems central to his approach. His scientific work has consistently been rooted in practical reality—a theory is valid to the extent that it actually can be made to work. If the theory doesn't work—and Hounsfield has been the first to admit that many of his ideas have not worked—he goes on to something that will work.

Current Biography quoted one of Hounsfield's colleagues as saying that while 99 of the ideas Hounsfield generates may not work, "the one hundredth is ten times better than anything anyone ever thought of before." In the same article, Hounsfield said, "The most important thing is a very broad understanding of the problem, not to be fogged by detail."

Hounsfield never married, and for a long time he didn't even establish a permanent residence. From his earliest years he enjoyed walking, particularly in England's Lake District. It was on one of these outings, which he termed "rambles," that he says he first got the idea for the X-ray scanner. His friends describe him, in the *Current Biography* article, as "courteous, soft-spoken, absentminded, and introverted." He is tall and slender, and his features are rather craggy. He wears a mustache. Indeed, he looks exactly the picture of an ideal English nobleman.

In his autobiography he explains, "As a bachelor, I have been able to devote a great deal of time to my general interest in science which more recently has included physics and biology. A great deal of my adult life has centered on my work. . . . Apart from my work, my greatest pleasures have been mainly out-of-doors, and although I no longer ski I greatly enjoy walking in the mountains and leading country rambles. I am fond of music, whether light or classical, and play the piano in a self-taught way. In company I enjoy lively way-out discussions."

He has used the money from the prizes he has been awarded to build a laboratory in the home he finally settled into. It is near the farm he grew up on, in Nottinghamshire.

"I'm not the sort of person to build model planes," he said. "I've always searched for original ideas; I am absolutely opposed to doing something someone else has done."

Godfrey Hounsfield Chronology

1919 Born in Newark, England, on August 28

1939 Attended City and Guilds College in London; volunteered for the Royal Air Force

1945 Awarded Certificate of Merit by RAF

1951 Received diploma in electrical and mechanical engineering from Faraday House Electrical Engineering College; began working for EMI, Ltd.

1958–59 Project engineer for the first English solid-state computer, the EMIDEC 1100

1967 Began working on the idea of computerized axial tomography

1971 First CAT scanner installed at Atkinson Morley's Hospital in Wimbledon, England

1972 Received MacRobert Award from the Fellowship of Engineering

1975 Received Albert Lasker Basic Medical Research Award

1976 Whole-body scanners produced; Hounsfield received Duddell Medal and Prize from the Institute of Physics and Gairdner Foundation International Award

1977 Became senior staff scientist at EMI

1978 Appointed professional fellow in imaging sciences at the University of Manchester

1979 Awarded the Nobel Prize in physiology or medicine, shared with Allan M. Cormack

1981 Knighted by the British government

Diagnostic Imaging Chronology

1895	Professor Wilhelm Roentgen discovers X rays
1901	Roentgen is awarded the first Nobel Prize in physics for the discovery of X rays
1906	X-ray contrast medium used for imaging kidneys
1910	Barium sulfate introduced as contrast medium for gastrointestinal diagnoses
1912	*Theory of Radioactivity* published by Marie Curie
1924	First radiographic images made of gall bladder, bile duct, and blood vessels
1945	First coronary artery imaging
1947	Research team at Bell Laboratories invents the transistor
1950	Nuclear medicine applied for imaging kidneys, heart, and skeletal system
1955	Panoramic X-ray images made of the entire jaw and teeth
1960	Ultrasound imaging developed to look at abdomen, kidneys, and fetuses
1970	X-ray mammography becomes widespread for imaging breasts
1971	Computerized axial tomography developed by British engineer Godfrey Hounsfield at EMI Central Research Laboratories
1972	Raymond Damadian applies for the first patent for scanning the human body by NMR
1974	Raymond Damadian receives patent for NMR scanner
1972	Dr. Paul Lauterbur develops nuclear magnetic resonance imaging (MRI)
1979	Nobel Prize in physiology or medicine awarded to Godfrey Hounsfield and Allan M. Cormack for the development of computer assisted tomography
1980	First MRI of the brain
1984	Three-dimensional image processing using CAT or MRI data

1985 Positron emission tomography (PET) developed at the University of California

1989 Spiral CAT allows fast volume scanning

1993 Open MRI systems developed for claustrophobic patients

Further Reading

Books and Periodicals

Aaseng, Nathan. *The Inventors: Nobel Prizes in Chemistry, Physics, and Medicine.* Minneapolis, Minn.: Lerner Publications, 1988.

Dibner, Bern. *Wilhelm Conrad Röntgen and the Discovery of X Rays.* New York: Franklin Watts, Inc., 1968.

Gherman, Beverly. *The Mysterious Rays of Dr. Röntgen.* New York: Atheneum, 1994.

"Hounsfield, Godfrey," *Current Biography*, April 1980.

Wasson, Tyler, (editor). *Nobel Prize Winners: An H. W. Wilson Biographical Dictionary.* New York: H. W. Wilson Co., 1987.

Internet Addresses

Allan Cormack Autobiography
www.nobel.se/medicine/laureates/1979/cormack-autobio.html

Godfrey Hounsfield Autobiography
www.nobel.se/medicine/laureates/1979/hounsfield -autobio.html

History and Descriptions of Diagnostic Imaging Procedures
www.imaginiscorp.com

"Hounsfield, Sir Godfrey Newbold" *Encyclopedia Britannica*
www.britannica.com

Interview with Wilhelm Roentgen in *McClure's Magazine*, May 1, 1986
www.cc.emory.edu.X-RAYS/century_09.htm

Glossary

contrast medium—a substance swallowed or injected for the purpose of providing stronger contrasts in diagnostic imaging.

Crookes tube—a vacuum tube used to control electrical current.

electromagnetic waves—waves of electric and magnetic force, including gamma rays, X rays, ultraviolet light, visible light, infrared light, and radio waves.

magnetic field—an area of force that forms in the space around anything that is magnetic or that carries an electric current.

particle physics—the study of the behavior of parts of atoms, such as protons and neutrons from the atom's nucleus.

photographic plate—before the development of celluloid film, the solid plate on which images were recorded by light.

radiation—a form of energy, such as electromagnetic waves and nuclear radiation, that is given off in waves or particles.

radioisotope—a radioactive form of an element.

radiology—the branch of medicine that uses radiation for diagnosis and treatment of disease.

tomography—the process of making thin slices, or sections, of X-ray images for maximum contrast of tissue densities.

wavelength—the distance between a certain point on one wave and a comparable point on the next wave.

X rays—electromagnetic waves of shorter wavelengths than visible light.

Index